S0-BND-377

A CLOSER LOOK BOOK

Published in the United States by
Gloucester Press in 1978

All rights reserved
Originated and designed by
David Cook and Associates
and produced by
The Archon Press Ltd
28 Percy Street
London W1P 9FF

First published in
Great Britain 1977 by
Hamish Hamilton
Children's Books Ltd
90 Great Russell Street
London WC1B 3PT

Dr. Bernard Stonehouse is
Chairman of the Postgraduate
School of Environmental Science,
University of Bradford

Printed in Italy
by Alfieri & Lacroix

Library of Congress Cataloging in Publication Data

Stonehouse, Bernard.
  A closer look at plant life.

  (A Closer look book)
  Includes index.
  SUMMARY: Describes the development of plant
life on earth and discusses types of plants and the
functions of their parts.
  1. Botany—Juvenile literature. [1. Botany] I. Orr,
Richard. II. Weare, Philip. III. Hincks, Gary. IV. Title.
QK49.S82   581   77-14978
ISBN 0-531-01430-4

A closer LOOK at PLANT LIFE

Bernard Stonehouse

Illustrated by
Richard Orr, Philip Weare, Gary Hincks

Gloucester Press · New York · 1978

Copyright © The Archon Press Ltd 1977

26732

The earth's green cover

We live in a green world. Wherever possible, plants cover the earth with a living, leaf-green blanket. Green forests, mountains, fields, swamps, and marshes make up a patchwork of plant life that covers the world. Lakes, rivers, and oceans, too, are green—filled with tiny plants too small to see. Only the very hot, very dry and very cold deserts of the natural world, and the artificial city-deserts of concrete, brick and dirty air stop the spread of plant life.

This green cover is older than humanity itself. It has covered the planet for over a billion years. Through long periods of mountain building and erosion, through the movement of the continents and the changing of climates, it has spread across the face of the earth. It has survived fires, floods, changes in sea level, and it has even improved, so that today the plant cover is richer in some ways than the one of 500 million years ago.

What is a plant? The tiny, one-celled diatom—one of thousands in a glass of pond water—is a plant. So is the tallest redwood in the California forests. So are sagebrush and mosses, water hyacinths and reeds in a lake, orange and gray lichens on a stone wall. Not all are green, but they all contain the green pigment chlorophyll—a substance that helps them to combine the energy of sunlight and the gases of the air to live. A few kinds of plants—bacteria and fungi, for example—manage without chlorophyll; but aside from plants no other form of life can use chlorophyll to make its daily food.

The family of plants

There are about 400,000 kinds
of plants in the world. Each
kind is called a species. The
similarities and differences
among the plants are used to
classify them, or divide them
into groups. In a small area,
like this woodland corner, many
kinds of plants can be found.
The simplest are the algae; next
are the mosses and liverworts;
then the club mosses, horsetails,
and ferns. The most advanced and
complicated green plants are
gymnosperms (which include
conifers, like the pine) and
angiosperms, or flowering plants,
which are the most numerous of all
plants. Blackberries and pansies
are in the undergrowth around
this oak tree (which is also a
flowering plant), together with
grasses and ferns; algae and
lichens cling to tree branches
in the shadier areas; fallen
leaves and fallen tree trunks
form a rich underlayer for
fungi and bacteria.

Life cycle

All animals, including human beings, need food to eat and oxygen-rich air to breathe. Only green plants can provide these. Plants take minerals from the earth, and a gas—carbon dioxide —from the air. They use energy from sunlight to change these ingredients into sugars and starches—food for growth. At the same time, they give out oxygen. Animals like the rabbit and caterpillar eat the plants, and may be eaten themselves by other animals. They "burn" oxygen in breathing, and exhale carbon dioxide. When plants and animals die, they fertilize the earth and feed new plants.

Photosynthesis

The yellow-flowered fleabane has roots (1) and a stem (2) containing tubes that run out into the veins of the leaves. Water and minerals are carried from the roots, through the tubes, to the leaves. Carbon dioxide is taken in through tiny openings on the undersides of the leaves. Using the sun's energy and the chlorophyll in the leaf cells, the plant changes carbon dioxide, minerals and water into sugar and oxygen.

Plant factories

The gases of the air and the energy of sunlight, plus water and minerals from the soil are the raw materials plants use to create the earth's green cover. The one-celled alga takes in sunlight and raw materials through its cell wall and creates what it needs inside its own tiny body. The large plant does the same thing, but it is like a factory, with separate departments. It takes in gases and sunlight through the leaves, minerals and water through the roots. The leaves are the laboratories where food is made.

The process of making sugars and starches from sunlight and air and releasing oxygen is called photosynthesis—the use of light in putting things together. It is performed by all plants that contain the green substance chlorophyll. But no animal can do it—not even humans. Plants store energy and materials for their own use and we take what we need from them. Plant eaters feed directly on plants; meat eaters get what they need indirectly—by eating plant eaters, or each other. Human beings and other omnivores eat and use both plants and animals. This is the way the living world works.

Have you thanked a green plant today? Plants can be thanked directly for cereal, beans and bread, and indirectly for bacon, eggs and milk. They can be thanked, too, for our air, for they give off oxygen during photosynthesis.

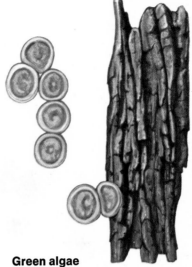

Green algae

Single-celled plants with no stems or leaves were probably the first plants to appear on the land. Many still exist. This *Protococcus* is an alga—a simple green plant→that lives on the bark of trees. Because of the oxygen they produced during photosynthesis, the earliest algae helped make it possible for life to develop on earth.

Fungi and bacteria

The microscopic bacteria and fungi (like molds and mushrooms) have no chlorophyll. They are often called decomposers. They break down dead plant and animal matter so it can in turn enrich the soil.

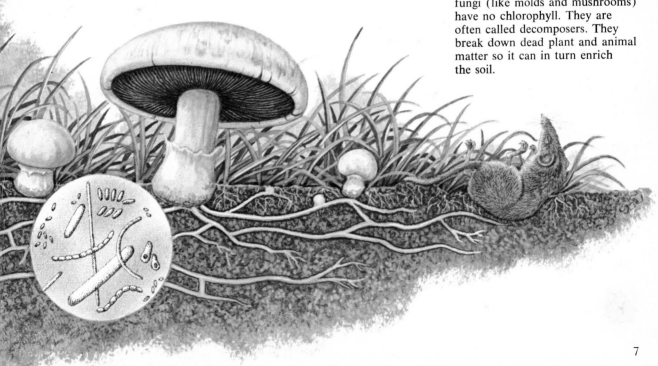

The unfriendly air

Life began in water over two billion years ago, and plants were established millions of years before real animals appeared. The first plants were probably like modern algae, and lived in the oceans like today's plankton and seaweed. Hundreds of millions of years passed before plants, and then animals, moved slowly onto land. Why did it take so long for plants to spread?

Water holds plants up so they don't need supporting structures. It keeps them at an even temperature and shields them from ultraviolet rays of the sun. It brings them dissolved minerals that can be taken in through cell walls. Above all, it keeps them from drying out.

Lichens and mosses
In this damp area, lichens grow on rocks and tree trunks where there is no soil. Algae cling to the bark higher up. Mosses grow in shallow soil at the edge of the stream and among roots.

The first land plants were probably pond plants that lasted through dry spells, or shore plants tough enough to survive from one tide to the next. But plants that followed had to adapt to land. Today's mosses, liverworts, and lichens show what these land-adapted plants may have been like.

Mosses and liverworts have no real roots and no systems to transport water. They take in water through their surfaces, but lose it easily and so must live in damp areas. Lichens are an odd combination of two plants—algae and fungi. The fungus part stores water and anchors the lichen to rock. The alga has chlorophyll and makes food. Lichens are tough and can survive drought and cold.

Moss reproduction
Moss plants (1) have male organs (2) and female organs (3). A drop of rain allows the male sperm (4) to swim across to the female. Then a cup or case forms (5). When it opens, special cells called spores (6) drift off and start new plants.

Relying on water

A damp home

Horsetails (below) usually grow in damp soil, with stems sometimes reaching up three feet (about 1 meter). The creeping club mosses (bottom) and ferns (far right) grow in dark, cool areas, often on the forest floor and among rocks at the edges of streams.

In the early days of the earth's history, low-growing mosslike plants competed for space and sunlight. A plant that could lift itself above its neighbors would get more space and light. To do this, the plants had to develop more complicated structures, with strong roots and stems for support. So, in time, some of the mosses changed, grew taller, and spread to drier areas. Among these were the earliest club mosses, horsetails, and ferns.

The important difference in structure between these plants and the mosses is the presence of steles—bundles of tubing that branch out through the roots, stems, and leaves. They carry water and food from one part of the plant to another, and they also strengthen the plant.

Descendants of these plants are only a very small part of the earth's cover today. There are still about a thousand species of club mosses, but most are tiny. Horsetails are bigger, but only about thirty species still exist. Ferns alone still thrive. There are over a thousand species, most in the temperate and tropical regions.

These early vascular plants (plants with systems of tubing) reproduced by means of spores (see fern diagram, right), and still needed damp living conditions to survive and reproduce. Later, new kinds of plants arose that reproduced by seeds, not spores, and so they no longer depended so much on water.

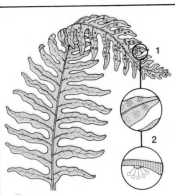

Ferns

The underside of a fern frond, or leaf (1) is covered with brown spore cases (2). When the cases open, the spores (3) fall onto damp ground. Each grows into a heart-shaped plant (a prothallus) about the size of a dime (4).

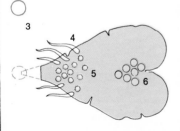

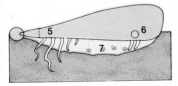

The plant develops male (5) and female (6) organs. Rain allows the male organs to release "sperm" (7) that swim to join an ovum made by the female organ, or ovary. The ovum develops into a new fern (8) that at first feeds on the prothallus.

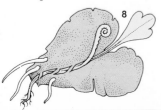

Seedless vascular plants

All these plants grow in areas where the ground is damp at least part of the year. In the large picture, from left to right, are: (1) horsetails; (2) staghorn club moss; (3) fir club moss; (4) wall rue; (5) rusty-back fern; (6) maidenhair spleenwort; (7) hart's-tongue fern.

The seed revolution

Cycads

Cycads, with their fernlike leaves and cones, show a half step in plant evolution between the early vascular plants and the seed plants.

Ferns, club mosses, and horsetails must live in damp places because they reproduce by means of spores that must have moisture for reproduction. These plants were able to grow only in certain areas of the earth. The next group of plants to develop were the seed producers. The evolution of the seed, about 400 million years ago, started a new trend of plant development. A seed is a package that contains a tiny, undeveloped plant (an embryo) and a supply of food—just enough to see the plant through the early stages of growth. The food supply and an outer covering—the seed coat—give the seed a chance to survive. Pollen and seeds allowed plants to spread from watery areas to all parts of the world. Today, the seed plants are the most numerous and important of all plants, and live everywhere except in the water itself, and in the harshest polar and mountain regions.

There are two kinds of seed plants, grouped according

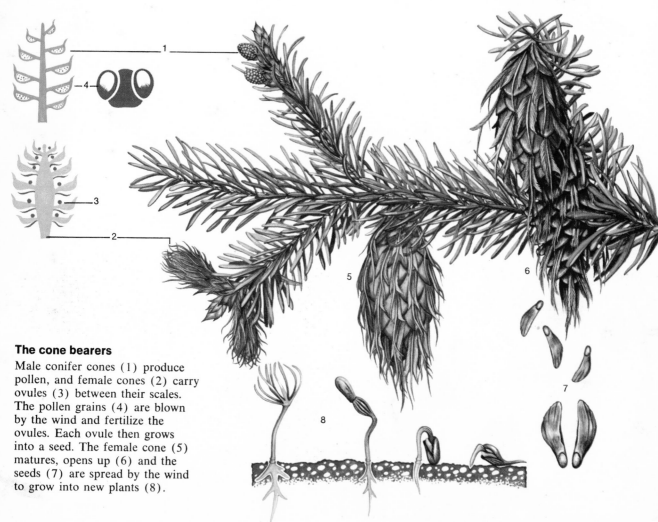

The cone bearers

Male conifer cones (1) produce pollen, and female cones (2) carry ovules (3) between their scales. The pollen grains (4) are blown by the wind and fertilize the ovules. Each ovule then grows into a seed. The female cone (5) matures, opens up (6) and the seeds (7) are spread by the wind to grow into new plants (8).

to the way the seeds are produced. The gymnosperms (having uncovered seeds) include the conifers. Conifer seeds lie naked, or uncovered, in their cones. The angiosperms (covered-seed plants) are the flowering plants, with seeds that are covered by fleshy fruit or a pod. There are many other differences between gymnosperms and angiosperms. Compare the pine on this page with the oak on page 27, for example.

Most of today's gymnosperms are conifers—pines, firs, spruces, cedars, yews, and other needle-leaved, cone-bearing trees that can grow in many areas—even where it is cold, windy, dry, or poor in soil. Their narrow leaves lose little moisture, so they can also grow in sandy, warm regions.

The evergreen conifer forests stretch almost around the top of the world, and long green fingers reach south along the mountain ranges of the northern continents.

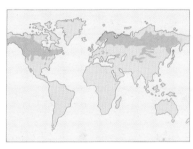

Conifers

The map shows the huge spread of conifer forests in the Northern Hemisphere. Conifers include the world's tallest trees, the 300-foot (90-meter) redwood (left); the oldest, bristlecone pines (below left), 4,000 years old; and many other species like the larch (below).

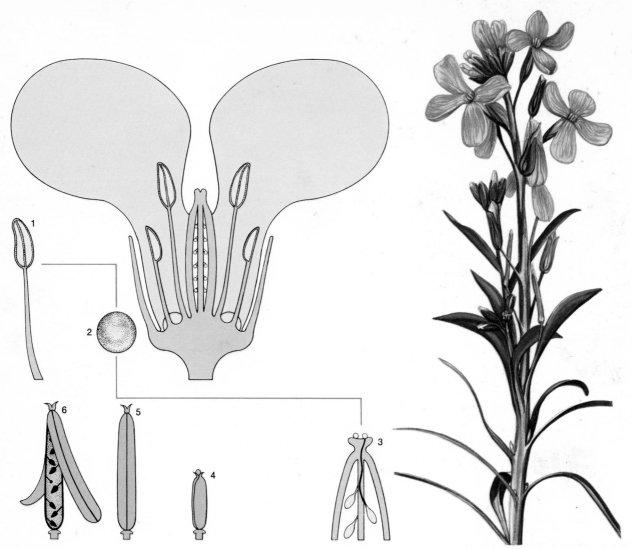

Flower structure

Flower petals surround
the stamens—the male reproductive
organs. And in the center of
the flower is the vase-shaped
pistil—the female organ. The
base of the pistil is the ovary
which produces the ovules, or
eggs, and the top is the stigma.
The stamens produce pollen. From
their tops, called the anthers (1),
pollen grains (2) are picked up
by wind or insects and carried
to the stigma (3) of another
flower. A tube grows down from
the pollen grain into the ovary
so the pollen can fertilize the
ovules (4). The ovules develop into
seeds and the ovary grows into
a fruit (5 and 6). When the fruit
is ripe, it bursts open and
the seeds are spread.

Spreading pollen

Many plants depend on insects to
carry the pollen needed to
fertilize the ovules. The flowers
attract the insects by smell,
pattern, and color. Some flowers,
at the tip of a single stem, are
too small to attract insects.
So they group together—with a
main stem and side branches
combining—to form clumps of
flowers. Wallflowers (above
right) attract insects by grouping
a number of flower heads on one
stem. On dandelions (right),
hundreds of florets are combined
on a single head. Crocuses
(far right) are examples of plants
which grow a single flower from
each bulb, but form clusters to
attract attention.

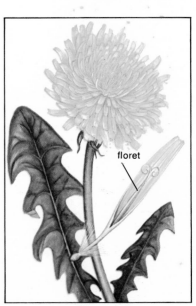

floret

Flowers

We don't know exactly how many kinds of flowering plants (angiosperms) exist, but there are more than a quarter of a million. New species are still being found, in fact may still be forming, for this is a fairly young group of plants and they may not have finished developing. It is hard to say when angiosperms first appeared, but hundreds of species already existed 150 million years ago.

Although younger, the flowering plants quickly caught up to and passed the gymnosperms, the first seed-producers. Angiosperms protect their seeds inside fleshy cases, giving them a better survival chance than the naked gymnosperm seeds. Also, gymnosperms depend on wind for pollination. Many flowers use wind, but more have developed special male and female organs (stamens and pistils) surrounded by bright petals that attract insects. Insect pollination is surer and less wasteful than wind pollination.

Flowering plants have other advantages. They are much more versatile—they can thrive in many more shapes and forms, and in more different situations than any other group. Most are soft-bodied herbs, others are trees, or woody shrubs, or vines. Rain forest trees are flowering plants; so is the desert cactus. Some live in water. The tropics are full of them and even the Arctic lands blaze with them for a short time each summer.

Holly
The holly produces male and female flowers on different trees. Berries are found only on those with female flowers.

Pollen

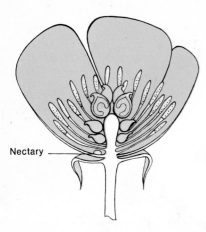

Nectar and pollen

This diagram of a buttercup shows the nectary—where nectar is produced—at the base of the petals. Insects searching for nectar have to brush against the pollen-loaded stamens.

In spite of their tiny size, the grains of pollen are living substances, produced by the male cones of conifers, or the male organs (anthers) of flowering plants. Each grain carries sperm cells to fertilize the ovules of the female organs. Plants that depend on wind pollination produce masses of very fine pollen that will stay in the air for several hours; pine pollen even has tiny air sacs to keep it afloat. Insect-pollinated plants produce less pollen, but it is sticky so that bees and other visitors will carry it from flower to flower.

Wind pollination is risky and wasteful, but it is still effective. It works well in forests, where there are hundreds of trees of the same species, and in grasslands where the wind is strong and millions of the same kind of plant ripen together. It is the only effective method for plants that flower in winter or early spring when few insects are about, or in polar and mountain regions where there are never many pollinating insects.

Insect pollination is usual among the angiosperms. Most flowers are formed so that insects, while collecting nectar, perform two vital tasks. One is to carry away some pollen from the flower they are visiting, the other is to leave pollen they have brought from other flowers of the same species. Many strangely shaped flowers are especially designed to make sure these tasks are carried out.

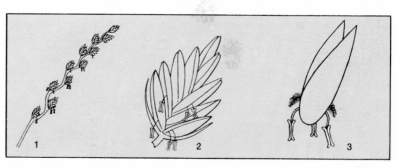

Wind-blown pollination

The rye grass (*left*) is a flowering plant. The stem is covered with tiny green flower clusters (1). Each cluster (2) is made up of individual flowers (3) whose feathery stigmas and stamens hang outside—waiting to catch the slightest breeze.

The insect contribution

Butterflies and bees, carrying pollen grains from flower to flower, provide plants with a sure method of pollination. After fertilization, the ovaries develop into fruit, so the work of insects is often a valuable service for agriculture.

Flowers for insects

Buttercups are easily pollinated when insects such as ants or beetles wander over them. The powerful, long-tongued bumblebee can force its way into a foxglove, but gets covered with pollen in the process. Masses of flowers of one species, like the clover, help to make sure that bees carry only one kind of pollen.

Fruits and seeds

Once a flower's ovules have been fertilized, its work is over. Sepals, petals, and stamens fall off, leaving the ovary and developing seeds. Food is rushed in from leaves and other organs where it has been stored. The seeds and ovary, and often the stalk tip, swell, ripen, and mellow into fruit.

Plants spread their seeds in two basic ways—by wind or by animal carriers. They make huge numbers of tiny seeds. Competition for living space means only a few survive.

Seeds that are spread by wind have special characteristics. Thistle and dandelion seeds float like little parachutes. Seeds of sycamore, ash, and many conifers are winged gliders that spin to earth. Some plants have seedpods that snap open and shoot out the seeds. Seeds that are spread by animals have hooks that catch onto fur or feathers. Others, which can survive digestion, are contained in fruits that attract hungry animals.

Wind dispersal
Each tiny dandelion seed is attached to a "parachute" of hairs that floats on the wind.

Animal dispersal
The seeds in the juicy blackberries (below) will be swallowed by the blackbird, and then pass through its stomach and be left somewhere else, where they will grow. Both plants and animals benefit from the process.

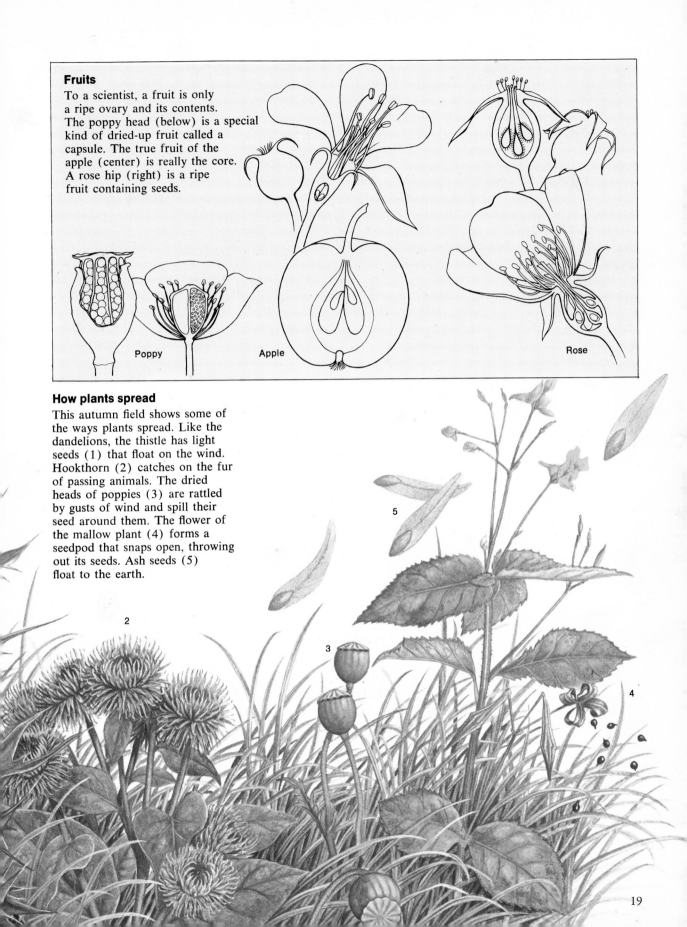

Fruits

To a scientist, a fruit is only a ripe ovary and its contents. The poppy head (below) is a special kind of dried-up fruit called a capsule. The true fruit of the apple (center) is really the core. A rose hip (right) is a ripe fruit containing seeds.

Poppy

Apple

Rose

How plants spread

This autumn field shows some of the ways plants spread. Like the dandelions, the thistle has light seeds (1) that float on the wind. Hookthorn (2) catches on the fur of passing animals. The dried heads of poppies (3) are rattled by gusts of wind and spill their seed around them. The flower of the mallow plant (4) forms a seedpod that snaps open, throwing out its seeds. Ash seeds (5) float to the earth.

19

Plant divisions

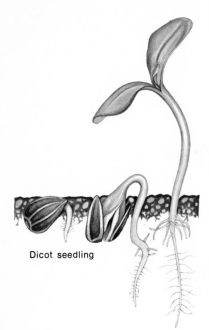

Dicot seedling

Human beings have always valued fruits and seeds. They taste better than most kinds of plant life and are more nourishing. Fruits are often rich in sugars and vitamin C, but are available only at certain seasons. Seeds can be eaten on the spot, or stored away. They are packed with starches, proteins, and oils. Our early ancestors, who were hunters and gatherers, may have realized the possibilities of farming when they saw stored grain sprout after a rain. Today, we manage the growing of billions of tons of cereal every year.

Seeds are made up of an embryo (a tiny, fully formed plant), complete with root (radicle), shoot (plumule), and one or two seed leaves (cotyledons) that hold the seed's food supply. These parts are packed into a tough, waterproof cover (testa) with a small hole. Although it is alive, the embryo can't germinate, or sprout, without water. With

Dicots

Dicots usually have rounded or divided leaves and petals arranged in fours or fives. The leaves have branching veins. The stem has a central core around which vascular bundles (the plant's "water pipes") are arranged in a circle.

Black-eyed Susan

Buttercup

Dandelion

water and the right environment, it starts to grow, and bursts out of its coat. The root unfolds, points downward, and grows into the soil. Tiny hairs appear around its tip, and the root begins to take in water from the soil, passing it upward into the growing leaf. The shoot grows upward, sometimes carrying the remains of the seed with it.

As the seedlings grow, we see that there are two kinds of flowering plants. The largest group is the class of dicotyledons, which have two seed leaves (cotyledons). We commonly call these dicots. The plants have rounded or divided leaves with branching veins that are not parallel.

Monocotyledons—monocots—form the other class of flowering plant. These have single seed-leaves; the shoot is one thin column. They have long, slender leaves with parallel veins. Grasses, irises, lilies, and palms are monocots.

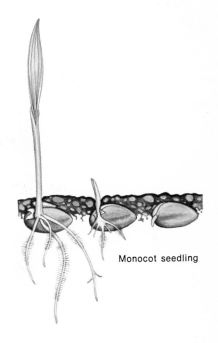

Monocot seedling

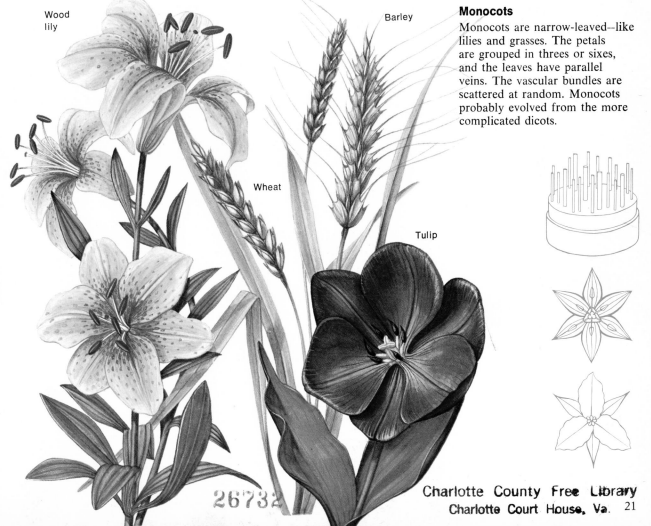

Wood lily

Barley

Wheat

Tulip

Monocots

Monocots are narrow-leaved--like lilies and grasses. The petals are grouped in threes or sixes, and the leaves have parallel veins. The vascular bundles are scattered at random. Monocots probably evolved from the more complicated dicots.

26732

Charlotte County Free Library
Charlotte Court House, Va.

Ways of living

Animals use legs, wings, and muscles to search for food. Plants stand still and grow roots. They stretch out their branches and take food from the air and sun. The advantage for plants is economy, in design and effort. But the plants are at the mercy of the climate—as you will realize if you've ever waited for a bus on a cold day. So plants arrange the cycles of their lives to fit, as closely as possible, the annual cycle of climate where they live.

Annuals are plants that complete their life cycle in one year or less. In temperate lands, many summer annuals germinate in spring, flower in summer, seed in autumn, and die back by winter. In deserts, the seeds of annuals may wait several years for rain. Then they sprout, and live only a few weeks.

Biennials take two growing years to complete their life cycle. The first summer they sprout, produce leaves and store food in stems, roots or leaf bases. The leaves may die in winter, but grow again in spring, using the stored food. By summer, the plant flowers and produces seed. Perennials continue from year to year. They include trees and shrubs and some softer-stemmed plants. They flower and make seed each year, and often grow much larger, too. Their main stems thicken and the branches grow longer.

Fireweed

Annuals

Annuals have no need for special food storage. They live for one growing year only. They sprout in spring, flower in summer, are fertilized and produce seeds (from which a new plant will grow the next year), and die in the fall.

Biennials

Biennials have a life cycle of two growing years. They sprout the first spring, grow leaves and store food to last through the winter. The leaves may die back but grow again the next spring. They produce flowers and then seeds in the second summer, and die that autumn.

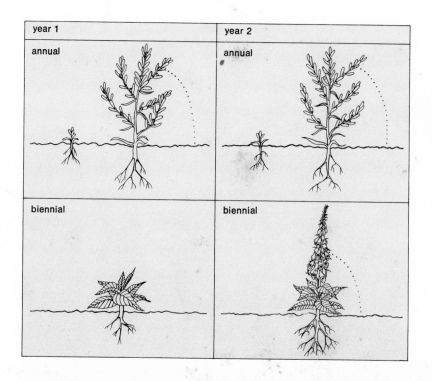

year 1   annual   year 2   annual   biennial   biennial

22

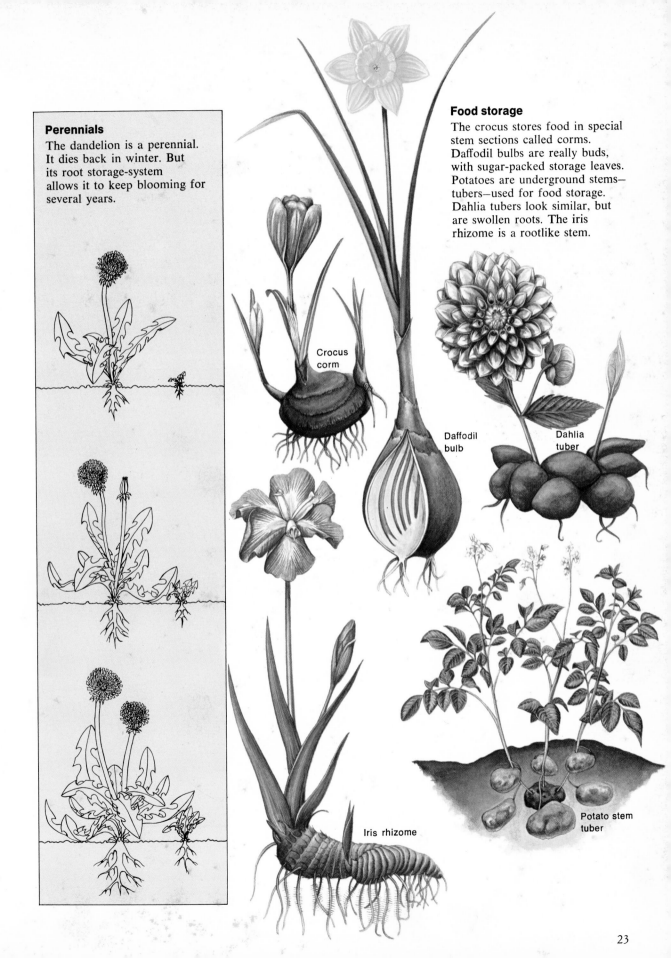

Perennials

The dandelion is a perennial. It dies back in winter. But its root storage-system allows it to keep blooming for several years.

Food storage

The crocus stores food in special stem sections called corms. Daffodil bulbs are really buds, with sugar-packed storage leaves. Potatoes are underground stems—tubers—used for food storage. Dahlia tubers look similar, but are swollen roots. The iris rhizome is a rootlike stem.

Crocus corm

Daffodil bulb

Dahlia tuber

Iris rhizome

Potato stem tuber

23

The woody plants

Chestnut twigs

The buds of some trees, like this chestnut, are really twigs, formed in the fall. The twig on the right is an autumn twig, with a terminal bud at the top and two side buds. The twig on the left shows two years' growth.

Annuals, biennials, and soft-stemmed perennials die back every autumn. Only woody perennials—trees and shrubs—last. Their strength lies in their wood—a chemical made in the walls of individual cells from starches and sugars. This tough springy material forms the bulk of the plant.

A tree grows upward and outward by making buds. The terminal bud (the one at the end of a twig) contains, in embryo form, the stem, side buds, and leaves of the new shoot. In spring, the cells in the bud divide again and again and expand, forcing open the covering leaves. Each length of stem between side buds grows, making the whole stem longer. Leaves and flowers unfold.

As the stem grows longer, its older parts thicken. Each year a new layer of wood is added under the bark, the outside layer. The tree thickens by adding a new ring of growth around its trunk and branches. There is a band of cells—the cambium—between the bark and the more solid wood of the trunk. On its outer side, the cambium manufactures phloem cells. These carry food from the leaves to all parts of the tree. On its inner side, the cambium makes xylem cells. These carry water and minerals from the roots to the tree. Xylem cells also contain the woody fibers that give strength and bulk to the tree trunk and branches.

Woodland layers

The tallest trees form an overhanging umbrella, or canopy layer. Trees and bushes in the shrub layer below spread out instead of up. They block light from the field layer of brambles and ferns below. The bottom layer is the woodland floor.

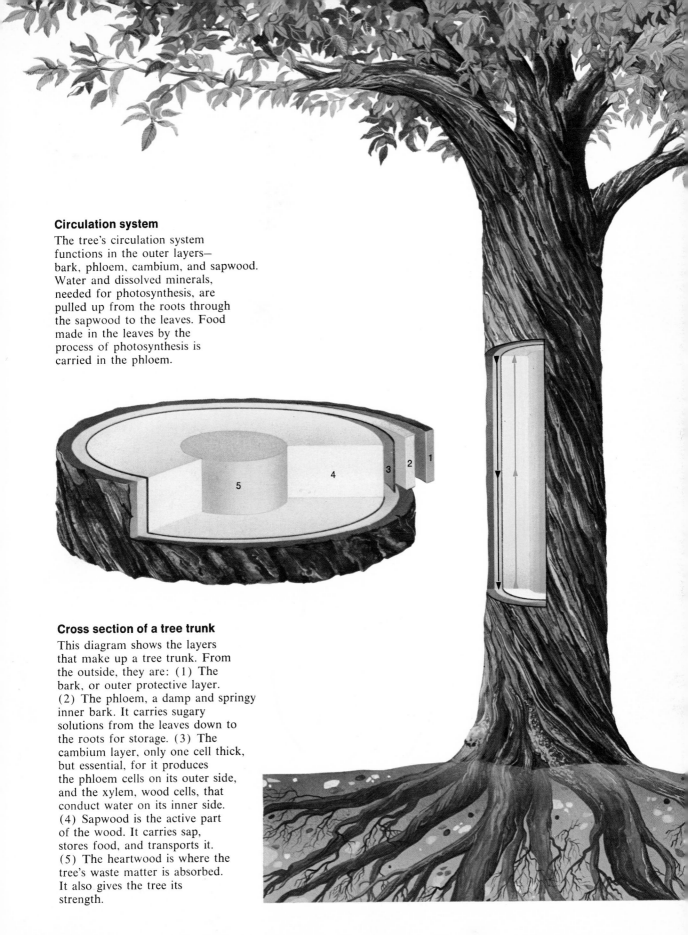

Circulation system

The tree's circulation system functions in the outer layers— bark, phloem, cambium, and sapwood. Water and dissolved minerals, needed for photosynthesis, are pulled up from the roots through the sapwood to the leaves. Food made in the leaves by the process of photosynthesis is carried in the phloem.

Cross section of a tree trunk

This diagram shows the layers that make up a tree trunk. From the outside, they are: (1) The bark, or outer protective layer. (2) The phloem, a damp and springy inner bark. It carries sugary solutions from the leaves down to the roots for storage. (3) The cambium layer, only one cell thick, but essential, for it produces the phloem cells on its outer side, and the xylem, wood cells, that conduct water on its inner side. (4) Sapwood is the active part of the wood. It carries sap, stores food, and transports it. (5) The heartwood is where the tree's waste matter is absorbed. It also gives the tree its strength.

Tree of life

A mature oak is more than a tree; it is a community center for many plants and animals who gather to live in, on, or around it. They all depend in one way or another on the oak and its products.

The tree, through photosynthesis, provides food for itself and many of its visitors. Some visitors—the ivy, and the algae, mosses, and lichens that cling to its bark—are photosynthesizers, too. They make their own food and use the tree only for support. Other creatures do take food, and are the first links in the many food chains that start from the oak. Thousands of insects feed on the tree, especially on the buds and fresh leaves of spring. Many of these are caterpillars that were laid as eggs on the branches in autumn, and hatch when the buds open. Grubs of sawflies, midges and gall wasps dig into the leaves and stems. They damage the tissues and cause lumps, or galls, on the surface. Weevils and aphids wander over the tree, drilling into leaves, twigs, and acorns, and sucking out the juices.

The insects have enemies. Aphids and other small insects fall prey to ladybug and lacewing larvae, and spiders. Larger insects are hunted by birds, especially titmice who nest in holes in the branches. Other birds nest in the leaves, in shrubs, or nearby holes. In autumn, larger birds, including bluejays, come to feed on the acorns, competing with squirrels and mice.

Woodland birds
The wood pigeon and the tawny owl, the green woodpecker (bottom right) and the nuthatch (bottom left) are common woodland dwellers.

Plants for all climates

There are three main types of vegetation, or plant life: forest, grassland, and desert. Climate is the factor that controls vegetation. Generally, wetter regions are forests, the drier tend to be grassland, and the driest—whether polar, temperate, or tropical—are desert, where few plants can grow.

The thickest vegetation is in the tropical rain forests of the hottest, wettest lands—from the Amazon River basin, to the islands of the South China Sea. Temperate forests grow in damper parts of temperate lands. The richest forests grow in warm coastal regions—California's redwood forests are an example. The coolest and driest of the world's great

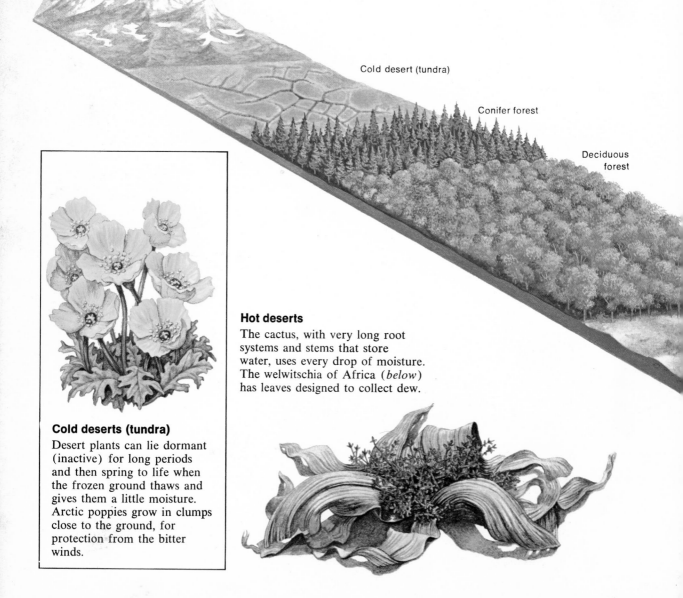

Cold desert (tundra)

Conifer forest

Deciduous forest

Cold deserts (tundra)
Desert plants can lie dormant (inactive) for long periods and then spring to life when the frozen ground thaws and gives them a little moisture. Arctic poppies grow in clumps close to the ground, for protection from the bitter winds.

Hot deserts
The cactus, with very long root systems and stems that store water, uses every drop of moisture. The welwitschia of Africa (*below*) has leaves designed to collect dew.

woodlands are the conifer forests of northern Europe, Asia, and North America.

Grasslands flourish where there is too little rain for forests. Where there is not enough rain even for grasses, the result is a desert. Tropical grasslands start at the drier edges of the tropical forests. Temperate grasslands border forests in cooler lands. Sometimes there are in-between areas of wooded parkland. Deserts are usually near grasslands, and may trade places with them over the centuries. The hottest, driest deserts are in Africa, the Middle East, and Australia. The coldest are in Greenland, Canada, and Antarctica.

Tropical forests
Tropical forests grow where the air is always warm and it rains every day. These forests have a greater variety of plant life than any other.

Tropical forest

Deciduous leaves
Deciduous trees are those that lose their leaves in winter. They need a period of cold, followed by spring warmth, to start new growth.

Hot desert

Conifer needles
Although conifers are found in temperate forests, they can also grow in poor soil which may be frozen or snow-covered half the year. Their sloping branches shed snow; their long-lasting needle leaves lose less water. Both conifers and deciduous trees have adapted themselves to cold winters and cannot live where it is hot all year.

Grassland

The enduring plants

Human life is short, and the life of most plants is shorter. But in the earth's long history, plants have outlived people by hundreds of millions of years. To compare how long plants and people have lived on earth, imagine the six billion years of the earth's history squeezed into a single year:

The first three or four months roll by without any life on earth, for the crust is still too hot and the atmosphere unformed. After a long period of cooling, the first plants appear in the sea, about mid-June. These are tiny, algalike fragments we find in rock three billion years old. Months more roll by. Water plants thrive, but it is not until early December that land plants appear. They spread quickly. By the second week of December, mosses, horsetails, ferns, and conifers cover the land, and the first four-footed animals browse among them. At Christmas, the flowering plants bloom and huge reptiles lumber through them. Birds and tiny mammals appear. There are no signs of humanlike creatures until late on December 30, when ape-ancestors start to walk upright. Human beings appear only during the last two or three hours of New Year's Eve.

Plants have lived long ages without us; we would not survive more than a moment without plants, and the bounty they give us every day of our lives.

Plant power

In the jungle at Angkor Wat in Cambodia, this giant head, spotted with lichens and overgrown by tree roots, is a reminder of the power of plants to survive—and outlive our attempts to conquer nature.

Index